Who Pooped?™
on the Colorado Plateau
Written by Gary D. Robson
Illustrated by Robert Rath
D0752072

FARCOUNTRY PRESS
HELENA, MONTANA

To my research assistant (and daughter), Heather.
- Gary

For Lucy and Thomas, my poop experts.
- Robert

ISBN: 978-1-56037-430-5

© 2008 Farcountry Press
Text © 2008 Gary D. Robson
Illustrations © 2008 Farcountry Press

Who Pooped? is a registered trademark of Farcountry Press.

All rights reserved. This book may not be reproduced in whole or in part by any means (with the exception of short quotes for the purpose of review) without the permission of the publisher.

For more information on our books,
write Farcountry Press, P.O. Box 5630, Helena, MT 59604;
call (800) 821-3874; or visit www.farcountrypress.com.

Book design by Robert Rath.

Manufactured by
Bang Printing
3323 Oak Street
Brainerd, MN 56401
in April 2017

 Produced and printed in the United States of America.

21 20 19 18 17 4 5 6 7 8

Library of Congress Cataloging-in-Publication Data

Robson, Gary D.
Who pooped on the Colorado Plateau? / written by Gary D. Robson ; Illustrated by Robert Rath.
p. cm.
ISBN-13: 978-1-56037-430-5 (pbk.)
ISBN-10: 1-56037-430-6 (pbk.)
1. Animal tracks--Colorado Plateau. I. Rath, Robert ill. II. Title.
QL768.R648 2008
591.979--dc22
2007030787

"Are we there yet?" Michael asked, as he squirmed in the back seat. "We've been driving for*ever*."

"We're already there, Michael," said Dad. "We've been on the Colorado Plateau for more than an hour. We'll be at the campsite by lunchtime."

"But we aren't in Colorado yet," said Michael's older sister, Emily.

"Most of the Colorado Plateau isn't in Colorado, Emily," Mom explained. "It's a flat area that covers parts of four states. It's high up in the mountains, too. We're more than a mile above sea level right now."

the STRAIGHT POOP

The Colorado Plateau is named for the red rocks you see all over it. "Colorado" is a Spanish word for "red colored."

"I'm tired of being in the car," said Michael.

"I'd like to stretch my legs, too," said Mom. "Let's go for a walk after we stop for lunch."

"Michael's too *scared* to walk in the woods," said Emily. "He's afraid he'll get eaten by a mountain lion." She curved her fingers like claws and growled.

"Stop it, Emily," said Mom, holding Michael's hand. "Nobody's getting eaten by anything."

Michael was excited about the trip, but Emily was right. He had just read a book about big cats and mountain lions, with their big teeth and claws, frightened him.

"I *am* kind of scared of mountain lions," Michael admitted.

"Don't worry," Dad told him. "Mountain lions are scared of people, too. We probably won't even see one."

"Besides, we'll learn all about mountain lions without ever getting close to one!" Mom said.

"How?" said Emily.

the STRAIGHT POOP

Never hike by yourself. Mountain lions almost never bother people hiking in groups.

"We can find out a lot about animals by what we find on the trail," said Mom.

"Let's look for some *sign*," Dad said. "We'll show you what Mom's talking about."

"Sign?" asked Michael, puzzled. "You mean like a sign at the zoo?"

Dad smiled. “A sign is a clue that an animal has left behind,” he said.

“See this?” said Mom. “See where the bark has been chewed off of those trees? That’s a sign a porcupine was having its lunch.”

the STRAIGHT POOP

Porcupines love to eat bark. Sometimes they eat the bark all the way around the tree, killing it. This is called “girdling.”

Michael forgot all about mountain lions and was excited about looking for clues. "Look! Footprints!" he said.

"You're right," said Mom. "Those are porcupine tracks. And look how it has dragged its tail in the middle."

the STRAIGHT POOP

It's best to leave porcupines alone. Their quills are very sharp and they can stick you if you get too close.

Dad pointed to something next to the tracks. It was lumpy, bumpy, and brown.

"This is porcupine scat," Dad said.

"*Scat?*" asked Emily. "What's scat?"

"It's a word hikers and trackers use for animal poop," Dad replied.

"See, Michael," said Dad. "We don't have to get up close to an animal to learn about it. Instead of a close encounter of the *scary* kind, we'll have a close encounter of the *poopy* kind."

Everybody laughed, and Mom made a gross-out face.

“Let’s see if we can find some more sign,” said Emily. “This is like a treasure hunt!”

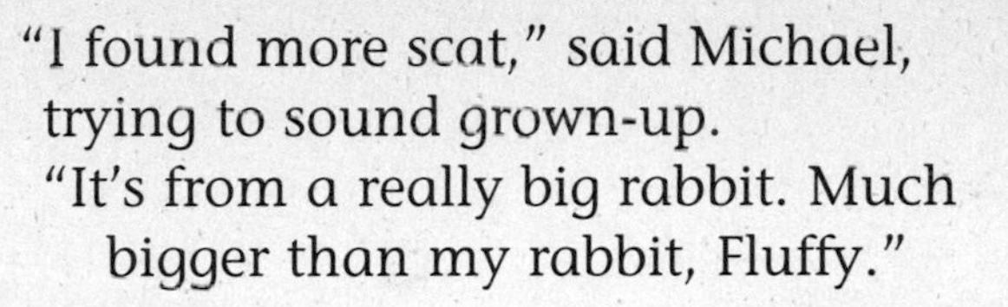

“I found more scat,” said Michael, trying to sound grown-up. “It’s from a really big rabbit. Much bigger than my rabbit, Fluffy.”

“Actually, Michael,” said Dad, “that’s deer scat.”

"How can you tell?" Emily asked.

"Rabbit scat is small and round, like little balls," Mom explained. "Deer scat is shaped more like jellybeans."

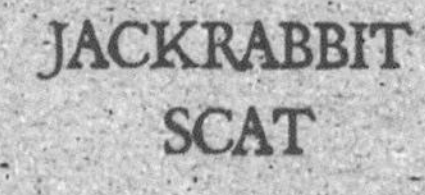

Michael spotted something on the ground and realized it was an antler. "Oh, no!" he said with a shiver. "Did a mountain lion eat the deer?"

"Don't worry, Michael," said Dad. "The deer shed their antlers every year and they grow bigger ones."

"This is a mule deer antler," said Mom. "You can tell because it is shaped differently than a white-tailed deer antler."

the STRAIGHT POOP

Horns are not the same as antlers. Antlers are shaped like branches and fall off every year. Horns never fall off and keep growing for the animal's entire life.

"Are these the deer's tracks over here?" asked Emily.

"Good eye!" said Mom. "See how the tracks are split down the middle? Deer hooves have two parts."

“This deer was in a hurry,” said Dad, as he studied the ground.

“How can you tell?” asked Emily.

"Over here, you can see the hoofprints get very far apart," Dad said. He pointed to the tracks. "And see how the front prints are behind the back prints?"

"Was the deer walking backwards?" asked Emily.

Mom chuckled. "No, it was galloping. Something scared it and it was moving fast," she said.

the STRAIGHT POOP
Sometimes mule deer bounce along with all four feet hitting the ground together. This is called "stotting" or "pronking."
RUNNING
FRONT HOOVES
BACK HOOVES
PRONKING
BACK HOOVES
FRONT HOOVES

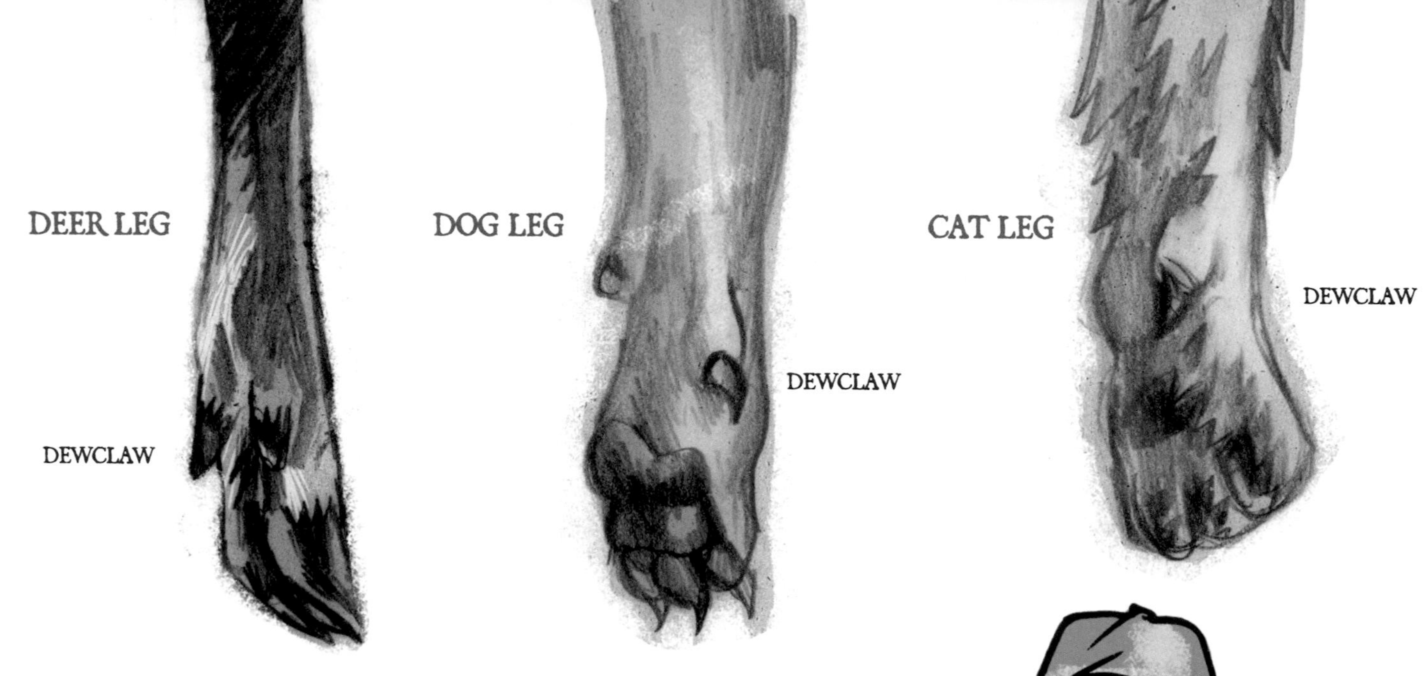

“What are these little marks?” asked Emily, pointing to two funny little dents behind each track.

“Those are from the deer’s *dewclaws,*” explained Mom.

Emily thought Mom was teasing her. “Deer don’t have claws!” she said.

“Dewclaws are small claws partway up an animal’s leg,” Mom said. “Lots of animals have them, including dogs and cats.”

“Here’s what scared the deer,” said Dad. “Look, there are coyote tracks and scat all around here.”

"They look like dog tracks," said Michael.

"That's because the coyote is part of the dog family," said Dad.

Mom added, "Their scat looks like dog poop, too, except that coyote scat has bits of bones and hair in it."

"Yuck!" said the kids.

"Some of these coyote tracks are very small," said Mom, "like they're from pups. There must be a den nearby. We should head out so we don't disturb them."

"Here's the rabbit scat you were looking for," said Dad.

"Yeah," said Michael, "but it is a lot bigger than Fluffy's!"

"I think this is jackrabbit scat," Dad said. "Jackrabbits are bigger than the desert cottontail rabbits around here—and so is their poop."

"You can't tell just from the poop what kind of rabbit or jackrabbit it was," Dad said. "The tracks show whether or not it is a rabbit or jackrabbit."

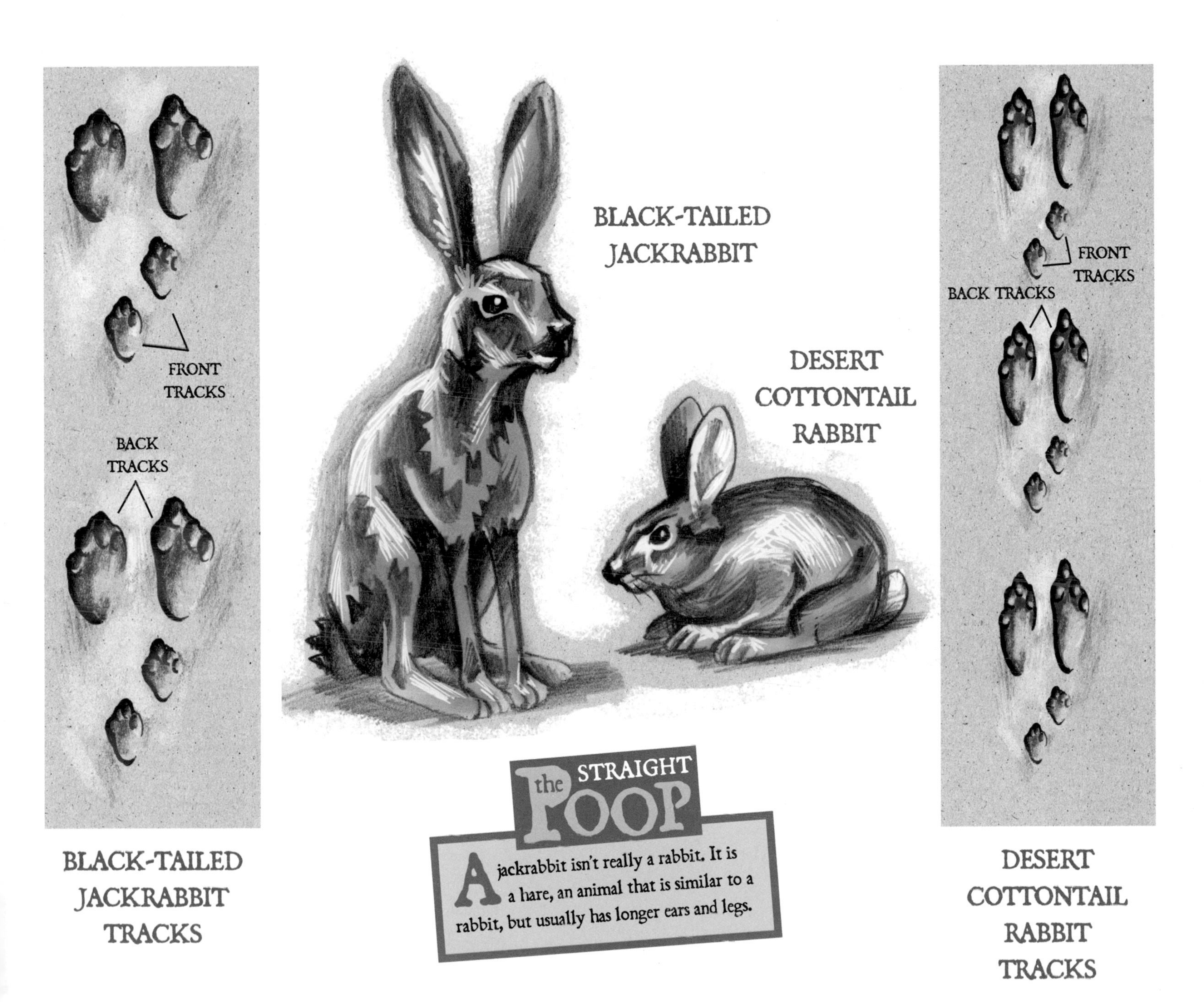

the STRAIGHT POOP

A jackrabbit isn't really a rabbit. It is a hare, an animal that is similar to a rabbit, but usually has longer ears and legs.

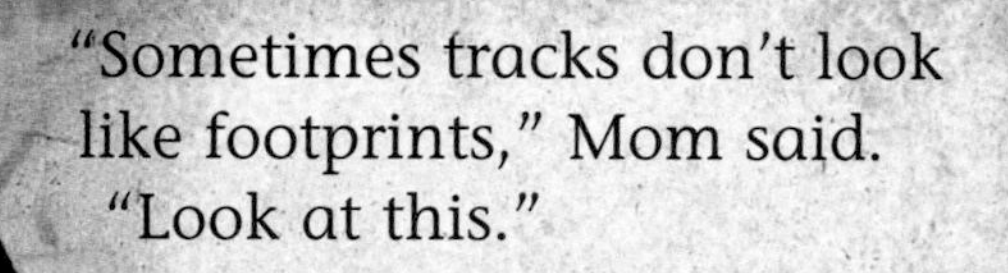

"Sometimes tracks don't look like footprints," Mom said. "Look at this."

"It looks like a squiggle pattern in the dirt," said Emily. "It's cool!"

"And here's its scat," said Mom.

"There's the animal that made the tracks—and the poop," said Mom.

"A rattlesnake!" yelled Michael.

"It's not a rattlesnake" said Dad. "It's a gopher snake, Michael. See his tail? There's no rattle on it. Farmers love gopher snakes because they eat a lot of rodents that can damage their crops."

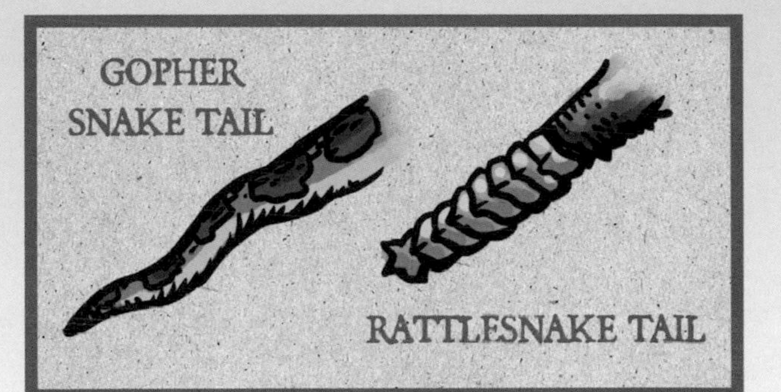

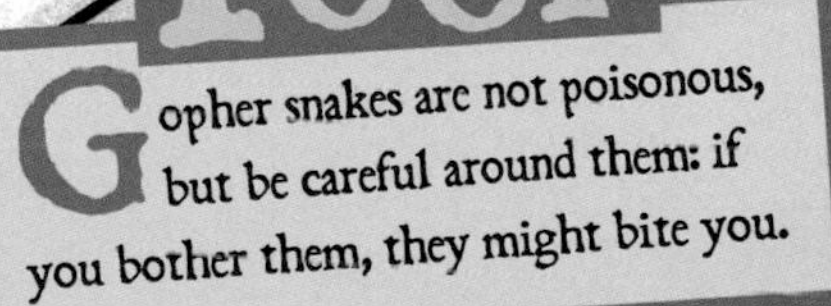

Gopher snakes are not poisonous, but be careful around them: if you bother them, they might bite you.

“Speaking of rodents,” Mom said, “can anyone guess who pooped over here?”

“A mouse?” guessed Michael. “The tracks are tiny.”

“It’s a ground squirrel,” said Dad. “Look, it’s sitting right over there.”

"How cute!" squealed Emily. "He looks like a chipmunk."

"Ground squirrels are close relatives of chipmunks," Mom told her.

the STRAIGHT POOP

Chipmunks, squirrels, mice, and rats are known as *rodents*. Rabbits, hares, and picas may look like rodents, but they're different types of animals called *lagomorphs*.

"Look at this!" Michael shouted. "What's *this* funny-looking poop?" He was excited about all of the scat they were finding.

"*Mii*-chael," Emily chimed in. "That's not poop. That's an owl pellet. We've been learning about them in school."

"That's close," said Dad. "It *is* a cough pellet, but if you look at these tracks next to it, you'll see it's from an eagle, not an owl."

the STRAIGHT POOP

An eagle track has three toes pointing forward and one pointing back, and it is much bigger than an owl's. An owl track has a pair of big toes in front, and two smaller toes behind or to the side.

"But what's a cough pellet?" Michael asked. "Does it mean the eagle has a cold?"

"Eagles and owls eat their prey whole—bones, fur, feathers, and all," Mom said. "What their tummies can't handle, they cough back up in pellets. If you pick the pellets apart, you can tell what the bird has been eating."

"A lot of people know about owl pellets, but don't realize that eagles have cough pellets, too," added Dad.

the STRAIGHT POOP

Golden eagles are closely related to America's national bird, the bald eagle.

“Is that a groundhog?” asked Emily.

“It’s a yellow-bellied marmot,” Dad answered. “A rodent that lives in high mountain areas.”

“Yellow-bellied marmot?” giggled Michael. “That’s fun to say!”

While the marmot and the kids looked at each other, Mom walked over to a group of small, twisted trees.

"Does anyone know what these tracks are?" she asked.

Michael joined her. “They look like the coyote tracks, only smaller.”

“You can’t see any claw marks,” said Emily.

“Good thinking,” said Mom. “These are from a gray fox. Their claws are so sharp and narrow that they often don’t appear in the tracks.”

the STRAIGHT POOP
Gray foxes are the only members of the dog family that can actually climb trees like cats.

"I found some *gigantic* gray fox footprints!" yelled Michael.

"Hmmm," said Mom. "These look like fox tracks, but they're not."

"You're right," said Emily. "These are bigger, there are no claw marks, and the front of the big pad looks dented in."

the STRAIGHT POOP

Since cats can retract their claws, their tracks usually don't show claw marks. All dogs, except gray foxes, almost always leave tracks with claw marks.

"You can also see where one toe sticks out farther in front than the others," said Mom. "It looks like you found your mountain lion, Michael."

"Where?" Michael said, turning pale.

"Sorry," said Mom. "I meant you found mountain lion *sign*."

"Now let's see what can you figure out about this cat," said Dad. "You kids do the detective work."

"I see a bunch of scratch marks on this tree," said Emily. "I think the mountain lion used it like a scratching post to sharpen its claws!"

the STRAIGHT POOP

Mountain lions have different names in different parts of the country. They're also known as catamounts, cougars, painters, panthers, and pumas.

"Is this mountain lion scat?" asked Michael.

"It sure is," said Dad. "See how it tried to bury its scat?"

"The scat has hair and bone in it, just like the coyote's," said Michael. "That means they eat other animals."

the STRAIGHT POOP

Mountain lions may be the biggest cat in America, but they still bury their scat just like a housecat.

Emily laid her hand next to the track.
"It's *really* big," she said.

"That's right," Mom said. "A mountain lion weighs as much as I do, and a big one can weigh more than Dad!"

As the family ate dinner that night, everyone talked about how much fun they were having on the Colorado Plateau.

"We didn't see very many animals," said Emily, "but it seemed like we did!"

Everyone laughed when Michael said, “And I didn’t get scared once!”

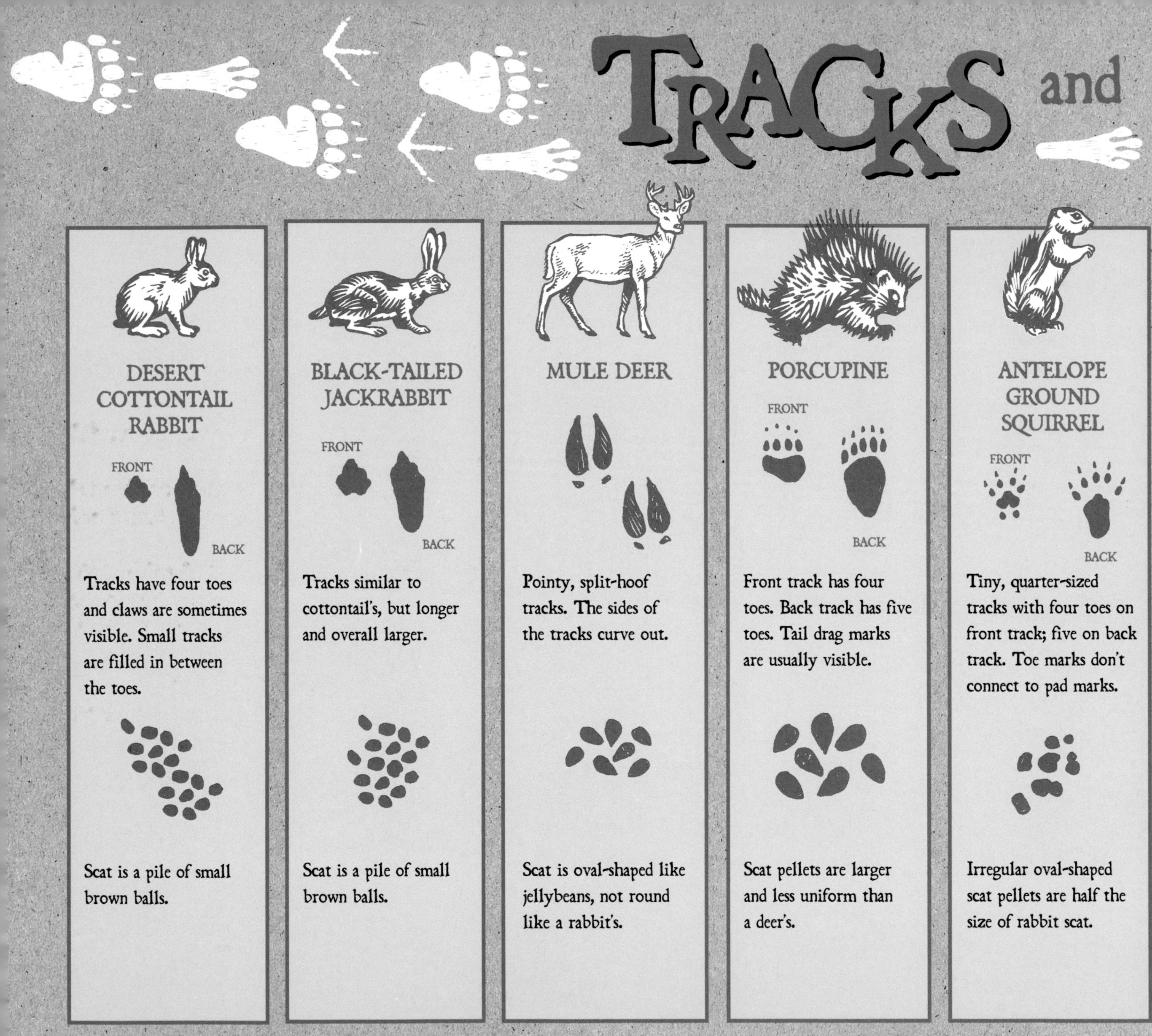
TRACKS and
DESERT COTTONTAIL RABBIT
FRONT
BACK
Tracks have four toes and claws are sometimes visible. Small tracks are filled in between the toes.
Scat is a pile of small brown balls.
BLACK-TAILED JACKRABBIT
FRONT
BACK
Tracks similar to cottontail's, but longer and overall larger.
Scat is a pile of small brown balls.
MULE DEER
Pointy, split-hoof tracks. The sides of the tracks curve out.
Scat is oval-shaped like jellybeans, not round like a rabbit's.
PORCUPINE
FRONT
BACK
Front track has four toes. Back track has five toes. Tail drag marks are usually visible.
Scat pellets are larger and less uniform than a deer's.
ANTELOPE GROUND SQUIRREL
FRONT
BACK
Tiny, quarter-sized tracks with four toes on front track; five on back track. Toe marks don't connect to pad marks.
Irregular oval-shaped scat pellets are half the size of rabbit scat.

SCAT NOTES

COYOTE

Track are like a dog's, with four toes and visible claw marks.

Scat is dark-colored, with tapered end, and usually contains hair.

MOUNTAIN LION

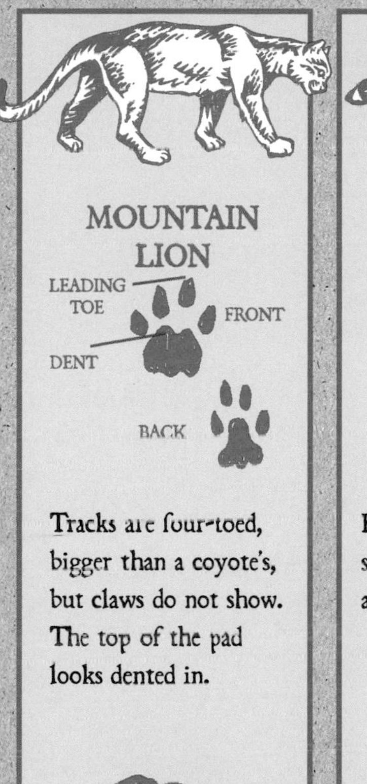

Tracks are four-toed, bigger than a coyote's, but claws do not show. The top of the pad looks dented in.

Scat is rarely seen because mountain lions bury it.

GRAY FOX

Four-toed tracks are smaller than a coyote's and claws are not visible.

Scat very similar to coyote's, but is often found on high places, such as rocks and tree branches.

GOPHER SNAKE

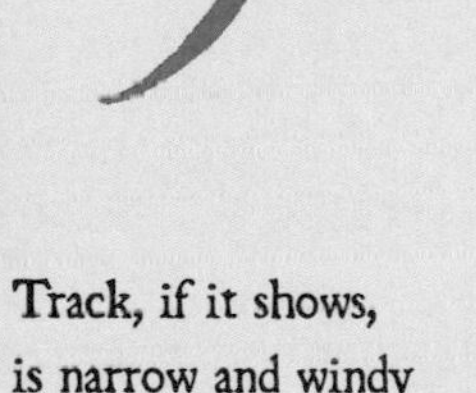

Track, if it shows, is narrow and windy or a smudge in the dirt.

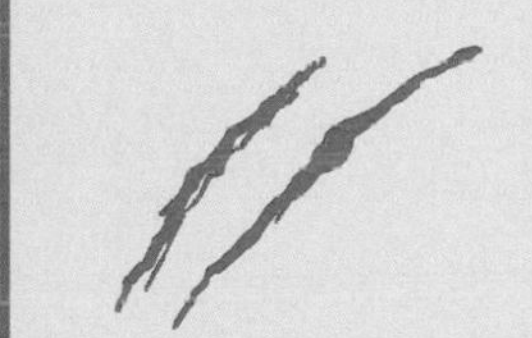

Scat is black or brown, long and stringy.

GOLDEN EAGLE

Tracks feature three slender toes pointing forward, and one toe pointing backward. Claw indentations do not touch the toe prints.

Scat is runny and white.

Cough pellets are over three inches long and contain hair.

ABOUT the AUTHOR and ILLUSTRATOR

GARY D. ROBSON

Gary Robson lives in Montana, not far from Yellowstone National Park. He has written dozens of books and hundreds of articles, mostly related to science, nature, and technology. www.GaryDRobson.com

ROBERT RATH

is a book designer and illustrator living in Bozeman, Montana. Although he has worked with Scholastic Books, Lucasfilm, and The History Channel, his favorite project is keeping up with his family.

Who Pooped?™

OTHER BOOKS IN THE
WHO POOPED?
SERIES:

Olympic
Glacier
Cascades
Black Hills
Northwoods
Acadi
Redwoods
Yellowstone
Grand Teton
Central Par
Yosemite
Rocky Mountain
Shenandoah
Sequoia/Kings Canyon
Death Valley
Red Rock Canyon
Grand Canyon
Colorado Plateau
Great Smoky Mountains
Sonoran Desert
Big Bend